FAITH OVER FEAR

MORE THAN A CONQUEROR
45-DAY DEVOTIONAL

JOURNAL

STEFANIE AUSTIN

FAITH OVER FEAR

MORE THAN A CONQUEROR
45-DAY DEVOTIONAL

JOURNAL

STEFANIE AUSTIN

Faith Over Fear – "More Than A Conqueror" - Journal

Request for information should be addressed to:
Stefanie.austin01@gmail.com

ISBN 978-1-7344974-9-6 (paperback)

Library of Congress Control Number:
2025922116

Electronic Version:
Amazon Kindle Publishing

Dedication

This book, Faith Over Fear: More Than a Conqueror, is lovingly dedicated to my mother, Amelia Morton.

Mom, you have been the foundation of my life; my greatest supporter, friend, cheerleader, disciplinarian, prayer warrior, and even a comedian when I needed laughter. You are not only a wonderful mother, grandmother, and great-grandmother, but also a friend to many and a mother to countless others whose lives you have touched. Your strength, humility, and unwavering faith have shaped me into who I am today.

I also honor my grandmothers and great-grandmothers; strong spiritual women who reminded us to never forget where we came from and to always remember who we belong to: our Heavenly Father, Jehovah.
From childhood, you instilled in me

the belief that I could do anything. Because of you, I learned to rise above ridicule, naysayers, and doubts, always pressing forward in faith. Though you are quiet and humble, your presence commands respect, and your wisdom reflects the depth of your life and faith. Coming from a home where the highest level of education was the sixth grade, you became a woman of knowledge, grace, and strength, living proof of God's power and favor.

I thank God for you, Mom. I thank Him for my father, for my family, my friends, and even strangers who have encouraged me along the way. But most of all, I want to give you your flowers now, while you can still receive them. You are my heart, and I love you deeply.

This book is for you mommy.

Introduction

Welcome to the Faith Over Fear - More Than A Conqueror Journal. This journal was prayerfully created as a companion to the devotional, offering you a quiet space to record your prayers and thoughts as you study God's Word.

Life brings moments that can stir fear, but Scripture reminds us that faith overcomes fear and that through Jesus Christ we have the strength to conquer every challenge.

As you reflect on these truths, use these pages to pour out your heart before God: your prayers, testimonies, and discoveries as you grow in faith. Allow this journal to become a record of God's faithfulness, showing how faith helps you rise above fear. You are invited to reflect on the marvelous works of God in your life and use this journal to celebrate gratitude, praise, and the many wins He's brought your way.

May these pages strengthen your walk with Christ and remind you that you are never alone - you are *More Than A Conqueror.*

2 Timothy 1:7 (KJV)

For God hath not given us the spirit of fear; but of power, and of love, and of a sound mind.

Prayers & Thoughts

Philippians 4:13 (KJV)

I can do all things through Christ which strengthens me.

Prayers & Thoughts

Psalm 55:22 (NKJV)

Cast your burden on the Lord,
And He shall sustain you;
He shall never permit the
righteous to be moved.

Prayers & Thoughts

Romans 8:37 (KJV)

Nay, in all these things we are more than conquerors through him that loved us.

Prayers & Thoughts

Proverbs 18:10 (NIV)

The name of the Lord is a
fortified tower;
the righteous run to it and
are safe.

Prayers & Thoughts

Isaiah 41:10 (KJV)

Fear thou not; for I am with thee: be not dismayed; for I am thy God: I will strengthen thee; yea, I will help thee; yea, I will uphold thee with the right hand of my righteousness.

Prayers & Thoughts

Joshua 1:9 (NKJV)

Have not I commanded thee? Be strong and of a good courage; be not afraid, neither be thou dismayed: for the Lord thy God is with thee whithersoever thou goest.

Prayers & Thoughts

Psalm 27:1 (KJV)

The Lord is my light and my salvation; whom shall I fear? the Lord is the strength of my life; of whom shall I be afraid?

Prayers & Thoughts

Psalm 34:4 (KJV)

I sought the Lord, and he heard me, and delivered me from all my fears.

Prayers & Thoughts

Psalm 46:1-2 (KJV)

God is our refuge and strength, a very present help in trouble. Therefore, will not we fear, though the earth be removed, and though the mountains be carried into the midst of the sea.

Prayers & Thoughts

Psalm 91:1-2 (KJV)

He that dwelleth in the secret place of the most High shall abide under the shadow of the Almighty. I will say of the Lord, He is my refuge and my fortress: my God; in him will I trust.

Prayers & Thoughts

Psalm 91:9-11 (KJV)

For he shall give his angels charge over thee, to keep thee in all thy ways.

Prayers & Thoughts

Nehemiah 8:10 (KJV)

"...for the joy of the Lord is your strength."

Prayers & Thoughts

Psalm 59:16 (NIV)

But I will sing of your strength,
in the morning I will sing of your love;
for you are my fortress,
my refuge in times of trouble.

Prayers & Thoughts

John 16:33 (KJV)

These things I have spoken unto you, that in me ye might have peace. In the world ye shall have tribulation: but be of good cheer; I have overcome the world.

Prayers & Thoughts

John 14:27 (KJV)

Peace I leave with you, my peace I give unto you: not as the world giveth, give I unto you. Let not your heart be troubled, neither let it be afraid.

Prayers & Thoughts

1 John 4:18 (KJV)

There is no fear in love; but perfect love casteth out fear: because fear hath torment. He that feareth is not made perfect in love.

Prayers & Thoughts

1 Corinthians 16:13 (KJV)

Watch ye, stand fast in the faith, quit you like men, be strong.

Prayers & Thoughts

Ephesians 6:10-11 (KJV)

Finally, my brethren, be strong in the Lord, and in the power of his might. Put on the whole armour of God...

Prayers & Thoughts

Isaiah 43:2 (KJV)

When thou passest through the waters, I will be with thee; and through the rivers, they shall not overflow thee: when thou walkest through the fire, thou shalt not be burned; neither shall the flame kindle upon thee.

Prayers & Thoughts

Isaiah 12:2 (KJV)

Behold, God is my salvation; I will trust, and not be afraid: for the Lord Jehovah is my strength and my song; he also is become my salvation.

Prayers & Thoughts

Isaiah 26:3 (KJV)

Thou wilt keep him in perfect peace, whose mind is stayed on thee: because he trusteth in thee.

Prayers & Thoughts

Proverbs 3:5-6 (KJV)

Trust in the Lord with all thine heart; and lean not unto thine own understanding. In all thy ways acknowledge him, and he shall direct thy paths.

Prayers & Thoughts

Mark 11:24 (KJV)

Therefore, I say unto you, What things soever ye desire, when ye pray, believe that ye receive them, and ye shall have them.

Prayers & Thoughts

Matthew 6:33-34 (KJV)

But seek ye first the kingdom of God, and his righteousness; and all these things shall be added unto you. Take therefore no thought for the morrow: for the morrow shall take thought for the things of itself. Sufficient unto the day is the evil thereof.

Prayers & Thoughts

Matthew 7:7-8 (KJV)

*Ask, and it shall be given you;
seek, and ye shall find; knock,
and it shall be opened unto you:
For every one that asketh
receiveth; and he that seeketh
findeth; and to him that knocketh
it shall be opened.*

Prayers & Thoughts

Hebrews 13:5-6 (NIV)

...for he hath said, I will never leave thee, nor forsake thee. So that we may boldly say, The Lord is my helper...

Prayers & Thoughts

Hebrews 11:1 (KJV)

Now faith is the substance of things hoped for, the evidence of things not seen.

Prayers & Thoughts

Hebrews 11:6 (KJV)

But without faith it is impossible to please him: for he that cometh to God must believe that he is, and that he is a rewarder of them that diligently seek him.

Prayers & Thoughts

James 1:2-3 (KJV)

My brethren, count it all joy when ye fall into divers temptations; Knowing this, that the trying of your faith worketh patience.

Prayers & Thoughts

James 1:5 (KJV)

If any of you lack wisdom, let him ask of God, that giveth to all men liberally, and upbraideth not; and it shall be given him.

Prayers & Thoughts

1 Peter 5:7 (KJV)

*Casting all your care upon him;
for he careth for you.*

Prayers & Thoughts

2 Corinthians 12:9 (KJV)

And he said unto me, My grace is sufficient for thee: for my strength is made perfect in weakness.

Prayers & Thoughts

2 Corinthians 5:7 (KJV)

For we walk by faith, not by sight.

Prayers & Thoughts

2 Corinthians 4:8-9 (KJV)

*We are troubled on every side,
yet not distressed; we are
perplexed, but not in despair;
Persecuted, but not forsaken; cast
down, but not destroyed;*

Prayers & Thoughts

2 Corinthians 4:16-18 (KJV)

For which cause we faint not; but though our outward man perish, yet the inward man is renewed day by day. For our light affliction, which is but for a moment, worketh for us a far more exceeding and eternal weight of glory; While we look not at the things which are seen, but at the things which are not seen: for the things which are seen are temporal; but the things which are not seen are eternal.

Prayers & Thoughts

Romans 12:12 (KJV)

Rejoicing in hope; patient in tribulation; continuing instant in prayer;

Prayers & Thoughts

Romans 15:13 (KJV)

Now the God of hope fill you with all joy and peace in believing, that ye may abound in hope, through the power of the Holy Ghost.

Prayers & Thoughts

Romans 8:28 (KJV)

And we know that all things work together for good to them that love God, to them who are the called according to his purpose.

Prayers & Thoughts

Romans 8:31 (KJV)

What shall we then say to these things? If God be for us, who can be against us?

Prayers & Thoughts

Psalm 150:6 (KJV)

Let everything that hath breath praise the Lord. Praise ye the Lord.

Prayers & Thoughts

Psalm 100:1-2 (KJV)

Make a joyful noise unto
the Lord, all ye lands.
Serve the Lord with gladness:
come before his presence with
singing

Prayers & Thoughts

Psalm 103:1-2 (KJV)

Bless the Lord, O my soul: and all that is within me, bless his holy name. Bless the Lord, O my soul, and forget not all his benefits:

Prayers & Thoughts

Psalm 95:1-2 (KJV)

O come, let us sing unto the Lord: let us make a joyful noise to the rock of our salvation. Let us come before his presence with thanksgiving, and make a joyful noise unto him with psalms.

Prayers & Thoughts

Psalm 98:1 (KJV)

O sing unto the Lord a new song; for he hath done marvelous things: his right hand, and his holy arm, hath gotten him the victory.

Prayers & Thoughts

Psalm 147:1 (KJV)

Praise ye the Lord: for it is good to sing praises unto our God; for it is pleasant; and praise is comely.

Prayers & Thoughts

Psalm 148:1-2 (KJV)

Praise ye the Lord. Praise ye the Lord from the heavens: praise him in the heights. Praise ye him, all his angels: praise ye him, all his hosts.

Prayers & Thoughts

Psalm 149:1 (KJV)

Praise ye the Lord. Sing unto the Lord a new song, and his praise in the congregation of saints.

Prayers & Thoughts

Psalm 150:1-2 (KJV)

Praise ye the Lord. Praise God in his sanctuary: praise him in the firmament of his power. Praise him for his mighty acts: praise him according to his excellent greatness.

Prayers & Thoughts

Luke 1:37 (KJV)

For with God nothing shall be impossible.

Prayers & Thoughts

Zephaniah 3:17 (KJV)

The Lord thy God in the midst of thee is mighty; he will save, he will rejoice over thee with joy; he will rest in his love, he will joy over thee with singing.

Prayers & Thoughts

Habakkuk 3:17-19 (NASB)

Even if the fig tree does not blossom, And there is no fruit on the vines,
If the yield of the olive fails,
And the fields produce no food,
Even if the flock disappears from the fold, And there are no cattle in the stalls,
Yet I will triumph in the LORD,
I will rejoice in the God of my salvation.
The Lord GOD is my strength,
And He has made my feet like deer's feet, And has me walk on my high places.

Prayers & Thoughts

Exodus 14:14 (KJV)

The Lord shall fight for you, and ye shall hold your peace.

Prayers & Thoughts

Exodus 15:2 (NIV)

The LORD is my strength and my defense; he has become my salvation.
He is my God, and I will praise him, my father's God, and I will exalt him.

Prayers & Thoughts

Deuteronomy 20:4 (NIV)

For the LORD your God is the one who goes with you to fight for you against your enemies to give you victory.

Prayers & Thoughts

1 Chronicles 16:34 (NASB)

Give thanks to the Lord, for He is good;
For His faithfulness is everlasting.

Prayers & Thoughts

2 Chronicles 20:15 (NASB)

and he said, "Listen, all you of Judah and the inhabitants of Jerusalem, and King Jehoshaphat: This is what the Lord says to you: 'Do not fear or be dismayed because of this great multitude, for the battle is not yours but God's.

Prayers & Thoughts

2 Chronicles 20:21 (NASB)

When he had consulted with the people, he appointed those who sang to the Lord and those who praised Him in holy attire, as they went out before the army and said, "Give thanks to the Lord, for His faithfulness is everlasting.

Prayers & Thoughts

1 Samuel 17:47 (KJV)

All those gathered here will know that it is not by sword or spear that the Lord saves; for the battle is the Lord's, and he will give all of you into our hands.

Prayers & Thoughts

Joshua 24:15 (NIV)

But if serving the Lord seems undesirable to you, then choose for yourselves this day whom you will serve, whether the gods your ancestors served beyond the Euphrates, or the gods of the Amorites, in whose land you are living. But as for me and my household, we will serve the Lord.

Prayers & Thoughts

Jeremiah 29:11 (KJV)

For I know the thoughts that I think toward you, saith the Lord, thoughts of peace, and not of evil, to give you an expected end.

Prayers & Thoughts

Lamentations 3:22-23 (KJV)

It is of the Lord's mercies that we are not consumed, because his compassions fail not. They are new every morning: great is thy faithfulness.

Prayers & Thoughts

Micah 7:7-8 (NIV)

*But as for me, I watch in
hope for the Lord, I wait for God
my Savior; my God will hear me.*

Israel Will Rise

*Do not gloat over me, my enemy!
Though I have fallen, I will
rise. Though I sit in darkness,
the Lord will be my light.*

Prayers & Thoughts

Nahum 1:7 (NASB)

The Lord is good,
A stronghold in the day of trouble,
And He knows those who take
refuge in Him.

Prayers & Thoughts

Psalm 40:1-3 (KJV)

I waited patiently for the Lord; and he inclined unto me, and heard my cry. He brought me up also out of a horrible pit, out of the miry clay, and set my feet upon a rock, and established my goings. And he hath put a new song in my mouth, even praise unto our God: many shall see it, and fear, and shall trust in the Lord.

Prayers & Thoughts

Psalm 42:11 (NIV)

Why, my soul, are you downcast? Why so disturbed within me? Put your hope in God, for I will yet praise him, my Savior and my God.

Prayers & Thoughts

Psalm 62:1-2 (NIV)

Truly my soul finds rest in God;
my salvation comes from him.
Truly he is my rock and my
salvation; he is my fortress; I will
never be shaken.

Prayers & Thoughts

Psalm 63:3-4 (NIV)

Because your love is better than
life, my lips will glorify you. I will
praise you as long as I live,
and in your name I will lift
up my hands.

Prayers & Thoughts

Psalm 84:11-12 (NIV)

For the Lord God is a sun and shield; the Lord bestows favor and honor; no good thing does he withhold from those whose walk is blameless. Lord Almighty, blessed is the one who trusts in you.

Prayers & Thoughts

Psalm 18:1-3 (NIV)

I love you, Lord, my strength. The Lord is my rock, my fortress and my deliverer; my God is my rock, in whom I take refuge, my shield and the horn of my salvation, my stronghold. I called to the Lord, who is worthy of praise, and I have been saved from my enemies.

Prayers & Thoughts

Psalm 23:1-3 (NKJV)

The Lord is my shepherd;
I shall not want. He makes me to
lie down in green pastures;
He leads me beside the still
waters. He restores my soul;
He leads me in the paths of
righteousness For His name's
sake.

Prayers & Thoughts

Psalm 27:3 (NKJV)

Though an army may encamp against me, My heart shall not fear; Though war may rise against me, In this I will be confident.

Prayers & Thoughts

Psalm 28:7 (NIV)

The Lord is my strength and my shield; my heart trusts in him, and he helps me. My heart leaps for joy, and with my song I praise him.

Prayers & Thoughts

Psalm 30:5 (NKJV)

For His anger *is but for* a
moment, His favor *is for* life;
Weeping may endure for a night,
But joy *comes* in the morning.

Prayers & Thoughts

Psalm 32:7 (NIV)

You are my hiding place;
you will protect me from
trouble and surround me with
songs of deliverance.

Prayers & Thoughts

Psalm 34:1 (NKJV)

I will bless the Lord at all times;
His
praise shall continually be in
my mouth

Prayers & Thoughts

Psalm 37:4-5 (NIV)

Take delight in the Lord,
and he will give you the desires of your heart.
Commit your way to the Lord;
trust in him and he will do this:

Prayers & Thoughts

Psalm 56:3-4 (NIV)

When I am afraid, I put my trust in you. In God, whose word I praise— in God I trust and am not afraid. What can mere mortals do to me?

Prayers & Thoughts

Proverbs 28:1 (NKJV)

The wicked flee when no one pursues, But the righteous are bold as a lion.

Prayers & Thoughts

Isaiah 40:29-31 (NIV)

He gives strength to the weary
and increases the power of the
weak.
Even youths grow tired and
weary, and young men stumble
and fall; but those who hope in
the LORD will renew their
strength.
They will soar on wings like
eagles; they will run and not
grow weary, they will walk and
not be faint.

Prayers & Thoughts

Isaiah 54:17 (NKJV)

No weapon formed against you
shall prosper, And every
tongue which rises against you
in judgment You shall condemn.
This is the heritage of the
servants of the LORD,
And their righteousness is from
Me," Says the LORD.

Prayers & Thoughts

Isaiah 55:12 (NIV)

You will go out in joy
and be led forth in peace;
the mountains and hills
will burst into song before you,
and all the trees of the field
will clap their hands.

Prayers & Thoughts

Isaiah 58:11 (NKJV)

The LORD will guide you continually, And satisfy your soul in drought, And strengthen your bones; You shall be like a watered garden, And like a spring of water, whose waters do not fail.

Prayers & Thoughts

Matthew 11:28-30 (NKJV)

Come to Me, all you who labor and are heavy laden, and I will give you rest. Take My yoke upon you and learn from Me, for I am gentle and lowly in heart, and you will find rest for your souls. For My yoke is easy and My burden is light.

Prayers & Thoughts

Matthew 19:26 (NASB)

And looking at them, Jesus said to them, With people this is impossible, but with God all things are possible.

Prayers & Thoughts

Mark 9:23 (ESV)

And Jesus said to him, "'If you can'! All things are possible for one who believes."

Prayers & Thoughts

Luke 10:19-20 (NIV)

I have given you authority to trample on snakes and scorpions and to overcome all the power of the enemy; nothing will harm you. However, do not rejoice that the spirits submit to you, but rejoice that your names are written in heaven.

Prayers & Thoughts

Luke 12:32 (NKJV)

Do not fear, little flock, for it is your Father's good pleasure to give you the kingdom.

Prayers & Thoughts

Luke 18:27 (ESV)

But he said, "What is impossible with man is possible with God."

Prayers & Thoughts

John 1:5 (NASB)

And the Light shines in the darkness, and the darkness did not grasp it.

Prayers & Thoughts

John 3:16 (NIV)

For God so loved the world that he gave his one and only Son, that whoever believes in him shall not perish but have eternal life.

Prayers & Thoughts

1 Corinthians 10:13 (KJV)

There hath no temptation taken you but such as is common to man: but God is faithful... will with the temptation also make a way to escape, that ye may be able to bear it.

Prayers & Thoughts

1 Corinthians 15:57 (KJV)

But thanks be to God, which giveth us the victory through our Lord Jesus Christ.

Prayers & Thoughts

Galatians 2:20 (KJV)

I am crucified with Christ: nevertheless I live; yet not I, but Christ liveth in me…

Prayers & Thoughts

Galatians 5:22-23 (KJV)

But the fruit of the Spirit is love,
joy, peace, longsuffering,
gentleness, goodness, faith,
meekness, temperance...

Prayers & Thoughts

Ephesians 3:20-21 (KJV)

Now unto him that is able to do exceedingly abundantly above all that we ask or think, according to the power that worketh in us,

Prayers & Thoughts

Philippians 4:4 (KJV)

Rejoice in the Lord always: and again I say, Rejoice.

Prayers & Thoughts

Philippians 4:6-7 (KJV)

Be careful for nothing; but in everything by prayer and supplication with thanksgiving let your requests be made known unto God. And the peace of God... shall keep your hearts and minds through Christ Jesus.

Prayers & Thoughts

Colossians 3:15 (KJV)

And let the peace of God rule in your hearts... and be ye thankful.

Prayers & Thoughts

1 Thessalonians 5:16-18 (KJV)

Rejoice evermore. Pray without ceasing. In everything give thanks...

Prayers & Thoughts

2 Thessalonians 3:16 (KJV)

Now the Lord of peace himself give you peace always by all means.

Prayers & Thoughts

1 Timothy 6:12 (KJV)

Fight the good fight of faith, lay hold on eternal life…

Prayers & Thoughts

2 Timothy 2:1 (KJV)

Thou therefore, my son, be strong in the grace that is in Christ Jesus.

Prayers & Thoughts

Hebrews 4:16 (KJV)

Let us therefore come boldly unto the throne of grace, that we may obtain mercy, and find grace to help in time of need.

Prayers & Thoughts

Hebrews 10:23 (KJV)

Let us hold fast the profession of our faith without wavering; (for he is faithful that promised;)

Prayers & Thoughts

James 4:7-8 (KJV)

Submit yourselves therefore to God. Resist the devil, and he will flee from you. Draw nigh to God, and he will draw nigh to you.

Prayers & Thoughts

1 Peter 1:3-5 (KJV)

Blessed be the God and Father of our Lord Jesus Christ, which according to his abundant mercy hath begotten us again unto a lively hope...

Prayers & Thoughts

1 Peter 2:9 (KJV)

But ye are a chosen generation, a royal priesthood, a holy nation, a peculiar people; that ye should shew forth the praises of him...

Prayers & Thoughts

Jude 1:24-25 (KJV)

Now unto him that is able to keep you from falling, and to present you faultless... To the only wise God our Saviour, be glory and majesty...

Prayers & Thoughts

Revelation 12:11 (KJV)

And they overcame him by the blood of the Lamb, and by the word of their testimony...

Prayers & Thoughts

Revelation 19:1 (KJV)

...I heard a great voice of much people in heaven, saying, Alleluia; Salvation, and glory, and honour, and power, unto the Lord our God:

Prayers & Thoughts

Revelation 22:17 (KJV)

And the Spirit and the bride say, Come. And let him that heareth say, Come. And let him that is athirst come. And whosoever will, let him take the water of life freely.

Prayers & Thoughts

Find Stefanie online at
www.stefanieaustininspires.com

On Facebook at
www.facebook.com/mywillorhiswill

Instagram
@STEFANIEAUSTININSPIRES

Testimonials

"An inspiring nudge to remain steadfast in faith, hold onto joy, and lean into God's direction."
Gloria Petersen, Author/Speaker
"This book ministered to my heart when I least expected it—I know it will minister to yours, too." **Jan Levin, Editor**
"Faith over Fear is a game changer that reminds us we are more than conquerors in Christ!"
Julie Jones - RN, Realtor

About the Author

Stefanie Austin is a bestselling author of *My Will or His Will – Learning to Live a Life to Surrender* and *My Will or His Will – Who Is Your GPS?* Her latest book, *Faith Over Fear – More Than a Conqueror,* reflects her passion for inspiring others through faith.

A servant leader, entrepreneur, philanthropist, and healthcare executive, Stefanie is a proud daughter, mother of three, and grandmother. She is blessed with a loving family, friends, and a supportive church family from various congregations for over twenty years. Rooted in God's love and the legacy of her grandparents, Stefanie encourages others to walk boldly in faith over fear.

Made in the USA
Coppell, TX
05 November 2026

11301606R10134